2019 SQA Specimen and Past Papers with Answers

National 5
MATHEMATICS

2018 & 2019 Exams
and 2017 Specimen Question Paper

HODDER
GIBSON
AN HACHETTE UK COMPANY

This book contains the official SQA 2018 and 2019 Exams, and the 2017 Specimen Question Paper for National 5 Mathematics, with associated SQA-approved answers modified from the official marking instructions that accompany the paper.

In addition the book contains study skills advice. This advice has been specially commissioned by Hodder Gibson, and has been written by experienced senior teachers and examiners in line with the new National 5 syllabus and assessment outlines. This is not SQA material but has been devised to provide further guidance for National 5 examinations.

Hodder Gibson is grateful to the copyright holders, as credited on the final page of the Answer section, for permission to use their material. Every effort has been made to trace the copyright holders and to obtain their permission for the use of copyright material. Hodder Gibson will be happy to receive information allowing us to rectify any error or omission in future editions.

Hachette UK's policy is to use papers that are natural, renewable and recyclable products and made from wood grown in sustainable forests. The logging and manufacturing processes are expected to conform to the environmental regulations of the country of origin.

Orders: please contact Bookpoint Ltd, 130 Park Drive, Milton Park, Abingdon, Oxon OX14 4SE. Telephone: (44) 01235 827827. Fax: (44) 01235 400454. Email education@bookpoint.co.uk Lines are open 9.00–5.00, Monday to Friday, with a 24-hour message answering service. Visit our website at www.hoddereducation.co.uk. If you have queries or questions that are not about an order, you can contact us at hoddergibson@hodder.co.uk

This collection first published in 2019 by
Hodder Gibson, an imprint of Hodder Education,
An Hachette UK Company
211 St Vincent Street
Glasgow G2 5QY

Typeset by Aptara, Inc.

Printed in the UK

A catalogue record for this title is available from the British Library

ISBN: 978-1-5104-7820-6

2 1

2020 2019

SCOTLAND EXCEL
We are an approved supplier on the Scotland Excel framework.
Schools can find us on their procurement system as:
Hodder & Stoughton Limited t/a Hodder Gibson.

Introduction

National 5 Mathematics

This book of SQA past papers contains the question papers used in the 2018 and 2019 exams (with answers at the back of the book). A specimen question paper reflecting the content and duration of the exam in 2018 is also included.

All of the question papers included in the book (2018, 2019 and the specimen question paper) provide excellent representative exam practice for the final exams. Using these papers as part of your revision will help you to develop the vital skills and techniques needed for the exam, and will help you to identify any knowledge gaps you may have.

It is always a very good idea to refer to SQA's website for the most up-to-date course specification documents. These are available at https://www.sqa.org.uk/sqa/47419

The course

The National 5 Mathematics course aims to enable you to develop the ability to:

- select and apply mathematical techniques in a variety of mathematical and real-life situations
- manipulate abstract terms in order to solve problems and to generalise
- interpret, communicate and manage information in mathematical form
- use mathematical language and explore mathematical ideas.

Before starting this course you should already have the knowledge, understanding and skills required to achieve a pass in National 4 Mathematics and/or be proficient in equivalent experiences and outcomes. This course enables you to further develop your knowledge, understanding and skills in algebra, geometry, trigonometry, numeracy, statistics and reasoning. The course content is summarised below.

Algebra	Geometry		Trigonometry
• Expanding brackets • Factorising • Completing the square • Algebraic fractions • Equation of straight line • Functional notation • Equations and inequations • Simultaneous equations • Change of subject of formulae • Graphs of quadratic functions • Quadratic equations and discriminant	• Gradient • Volume • Properties of shapes • Vectors and 3D coordinates	• Arc and sector of circle • Pythagoras' theorem • Similarity	• Graphs • Equations • Identities • Area of triangle, sine rule, cosine rule, bearings
	Numeracy		**Statistics**
	• Surds • Indices and scientific notation • Rounding (significant figures) • Percentages • Fractions		• Semi-interquartile range, standard deviation • Scattergraphs; equation of line of best fit
Reasoning			
• Interpreting a situation where mathematics can be used and identifying a strategy. • Explaining a solution and/or relating it to context.			

Assessment

The course assessment is an examination comprising two question papers.

The number of marks and the times allotted for the examination papers are as follows:

Paper 1 (non-calculator)	50 marks	1 hour 15 minutes
Paper 2	60 marks	1 hour 50 minutes

The course assessment is graded A-D, the grade being determined by the total mark you score in the examination.

Some tips for achieving a good mark

- **DOING** maths questions is the most effective use of your study time. You will benefit much more from spending 30 minutes doing maths questions than spending several hours copying out notes or reading a maths textbook.
- Practise doing the type of questions that are likely to appear in the exam. Work through these practice papers and similar questions from past Credit Level and Intermediate 2 papers. Use the marking instructions to check your answers and to understand what the examiners are looking for. Ask your teacher for help if you get stuck.
- **SHOW ALL WORKING CLEARLY.** The instructions on the front of the exam paper state that "To earn full marks you must show your working in your answers". A "correct" answer with no working may only be awarded partial marks or even no marks at all. An incomplete answer will be awarded marks for any appropriate working. Attempt every question, even if you are not sure whether you are correct or not. Your solution may contain working which will gain some marks. A blank response is certain to be awarded no marks. Never score out working unless you have something better to replace it with.

- Communication is very important in presenting solutions to questions. Diagrams are often a good way of conveying information and enabling markers to understand your working. Where a diagram is included in a question, it is often good practice to copy it and show the results of your working on the copy.
- In Paper 1, you have to carry out calculations without a calculator. Candidates' performance in number skills is often disappointing, and costs many of them valuable marks. Ensure that you practise your number skills regularly, especially within questions testing Course content.
- In Paper 2, you will be allowed to use a calculator. Always use **your own** calculator. Different calculators often function in slightly different ways, so make sure that you know how to operate yours. Having to use a calculator that you are unfamiliar with on the day of the exam will disadvantage you.
- Prepare thoroughly to tackle questions from **all** parts of the course. Numerical and algebraic fractions, graphs of quadratic fractions, surds, indices and trigonometric identities are topics that often cause candidates problems. Be prepared to put extra effort into mastering these topics.

Some common errors to avoid

	Common error	Correct answer
Converse of Pythagoras' Theorem e.g. Prove that triangle ABC is right angled. 	Don't start by assuming what you are trying to prove is true. $AC^2 = AB^2 + BC^2$ $AC^2 = 3^2 + 4^2 = 9 + 16 = 25$ $AC = \sqrt{25} = 5$ so triangle ABC is right angled by the Converse of Pythagoras' Theorem.	Don't state that $AC^2 = AB^2 + BC^2$ until you have the evidence to prove that it is true. $AC^2 = 5^2 = 25$ $AB^2 + BC^2 = 3^2 + 4^2 = 9 + 16 = 25$ so $AC^2 = AB^2 + BC^2$ so triangle ABC is right angled by the Converse of Pythagoras' Theorem.
Similarity (area and volume) e.g. Theses cylinders are mathematically similar. The volume of the small one is $60cm^3$. Calculate the volume of the large one.	Don't use the linear scale factor to calculate the volume (or area) of a similar shape. Scale factor = 2 Volume = $2 \times 60 = 120cm^3$	Remember that volume factor = (linear factor)^3 area factor = (linear factor)^2 Scale factor = 2 Volume = $2^3 \times 60 = 480cm^3$
Reverse use of percentage e.g. After a 5% pay rise, Ann now earns £252 per week. Calculate her weekly pay before the rise.	Increase = 5% of old pay **NOT** 5% of new pay Increase = 5% of £252 = £12·60 Old pay = £252 - £12·60 = £239·40	New pay = (100% + 5%) of old pay New pay = 105% of old pay = £252 1% of old pay = £252 ÷ 105 = £2·40 Old pay = 100% = £2·40 × 100 = £240
Interpreting statistics e.g. Jack and Jill sat tests in the same eight subjects. Jack's mean mark was 76 and his standard deviation was 13. Jill's mean mark was 59 and her standard deviation was 21. Make two valid comments comparing the performance of Jack and Jill in the tests.	This answer does not show that you **understand** the meaning of mean and standard deviation. Jack has a higher mean mark but a lower standard deviation than Jill.	Your interpretation of the figures must show that you **understand** that mean is an average and that standard deviation is a measure of spread. On average Jack performed better than Jill as his mean mark was higher. Jack's performance was more consistent than Jill's as the standard deviation of his marks was lower.

Good luck!

Remember that the rewards for passing National 5 Mathematics are well worth it! Your pass will help you get the future you want for yourself. In the exam, be confident in your own ability, if you're not sure how to answer a question trust your instincts and just give it a go anyway – keep calm and don't panic! GOOD LUCK!

Study Skills – what you need to know to pass exams!

General exam revision: 20 top tips

When preparing for exams, it is easy to feel unsure of where to start or how to revise. This guide to general exam revision provides a good starting place, and, as these are very general tips, they can be applied to all your exams.

1. Start revising in good time.

Don't leave revision until the last minute – this will make you panic and it will be difficult to learn. Make a revision timetable that counts down the weeks to go.

2. Work to a study plan.

Set up sessions of work spread through the weeks ahead. Make sure each session has a focus and a clear purpose. What will you study, when and why? Be realistic about what you can achieve in each session, and don't be afraid to adjust your plans as needed.

3. Make sure you know exactly when your exams are.

Get your exam dates from the SQA website and use the timetable builder tool to create your own exam schedule. You will also get a personalised timetable from your school, but this might not be until close to the exam period.

4. Make sure that you know the topics that make up each course.

Studying is easier if material is in manageable chunks – why not use the SQA topic headings or create your own from your class notes? Ask your teacher for help on this if you are not sure.

5. Break the chunks up into even smaller bits.

The small chunks should be easier to cope with. Remember that they fit together to make larger ideas. Even the process of chunking down will help!

6. Ask yourself these key questions for each course:

- Are all topics compulsory or are there choices?
- Which topics seem to come up time and time again?
- Which topics are your strongest and which are your weakest?

Use your answers to these questions to work out how much time you will need to spend revising each topic.

7. Make sure you know what to expect in the exam.

The subject-specific introduction to this book will help with this. Make sure you can answer these questions:

- How is the paper structured?
- How much time is there for each part of the exam?
- What types of question are involved? These will vary depending on the subject so read the subject-specific section carefully.

8. Past papers are a vital revision tool!

Use past papers to support your revision wherever possible. This book contains the answers and mark schemes too – refer to these carefully when checking your work. Using the mark scheme is useful; even if you don't manage to get all the marks available first time when you first practise, it helps you identify how to extend and develop your answers to get more marks next time – and of course, in the real exam.

9. Use study methods that work well for you.

People study and learn in different ways. Reading and looking at diagrams suits some students. Others prefer to listen and hear material – what about reading out loud or getting a friend or family member to do this for you? You could also record and play back material.

10. There are three tried and tested ways to make material stick in your long-term memory:

- Practising – e.g. rehearsal, repeating
- Organising – e.g. making drawings, lists, diagrams, tables, memory aids
- Elaborating – e.g. incorporating the material into a story or an imagined journey

11. Learn actively.

Most people prefer to learn actively – for example, making notes, highlighting, redrawing and redrafting, making up memory aids, or writing past paper answers. A good way to stay engaged and inspired is to mix and match these methods – find the combination that best suits you. This is likely to vary depending on the topic or subject.

12. Be an expert.

Be sure to have a few areas in which you feel you are an expert. This often works because at least some of them will come up, which can boost confidence.

13. Try some visual methods.

Use symbols, diagrams, charts, flashcards, post-it notes etc. Don't forget – the brain takes in chunked images more easily than loads of text.

14. Remember – practice makes perfect.

Work on difficult areas again and again. Look and read – then test yourself. You cannot do this too much.

15. Try past papers against the clock.

Practise writing answers in a set time. This is a good habit from the start but is especially important when you get closer to exam time.

16. Collaborate with friends.

Test each other and talk about the material – this can really help. Two brains are better than one! It is amazing how talking about a problem can help you solve it.

17. Know your weaknesses.

Ask your teacher for help to identify what you don't know. Try to do this as early as possible. If you are having trouble, it is probably with a difficult topic, so your teacher will already be aware of this – most students will find it tough.

18. Have your materials organised and ready.

Know what is needed for each exam:
- Do you need a calculator or a ruler?
- Should you have pencils as well as pens?
- Will you need water or paper tissues?

19. Make full use of school resources.

Find out what support is on offer:
- Are there study classes available?
- When is the library open?
- When is the best time to ask for extra help?
- Can you borrow textbooks, study guides, past papers, etc.?
- Is school open for Easter revision?

20. Keep fit and healthy!

Try to stick to a routine as much as possible, including with sleep. If you are tired, sluggish or dehydrated, it is difficult to see how concentration is even possible. Combine study with relaxation, drink plenty of water, eat sensibly, and get fresh air and exercise – all these things will help more than you could imagine. Good luck!

NATIONAL 5

2017 Specimen
Question Paper

Mark

N5

National
Qualifications
SPECIMEN ONLY

S847/75/01

Mathematics
Paper 1
(Non-Calculator)

Date — Not applicable

Duration — 1 hour 15 minutes

Fill in these boxes and read what is printed below.

Full name of centre

Town

Forename(s)

Surname

Number of seat

Date of birth

| Day | Month | Year | Scottish candidate number |

Total marks — 50

Attempt ALL questions.

You may NOT use a calculator.

To earn full marks you must show your working in your answers.

State the units for your answer where appropriate.

Write your answers clearly in the spaces provided in this booklet. Additional space for answers is provided at the end of this booklet. If you use this space you must clearly identify the question number you are attempting.

Use **blue** or **black** ink.

Before leaving the examination room you must give this booklet to the Invigilator; if you do not, you may lose all the marks for this paper.

FORMULAE LIST

The roots of $ax^2 + bx + c = 0$ are $x = \dfrac{-b \pm \sqrt{(b^2 - 4ac)}}{2a}$

Sine rule: $\dfrac{a}{\sin A} = \dfrac{b}{\sin B} = \dfrac{c}{\sin C}$

Cosine rule: $a^2 = b^2 + c^2 - 2bc\cos A$ or $\cos A = \dfrac{b^2 + c^2 - a^2}{2bc}$

Area of a triangle: $A = \frac{1}{2}ab\sin C$

Volume of a sphere: $V = \frac{4}{3}\pi r^3$

Volume of a cone: $V = \frac{1}{3}\pi r^2 h$

Volume of a pyramid: $V = \frac{1}{3}Ah$

Standard deviation: $s = \sqrt{\dfrac{\Sigma(x - \bar{x})^2}{n-1}}$

or $s = \sqrt{\dfrac{\Sigma x^2 - \dfrac{(\Sigma x)^2}{n}}{n-1}}$, where n is the sample size.

MARKS | DO NOT WRITE IN THIS MARGIN

Total marks — 50

Attempt ALL questions

1. Evaluate

$$2\frac{3}{8} \div \frac{5}{16}.$$

2

2. Solve algebraically the inequality

$$11 - 2(1 + 3x) < 39.$$

3

[Turn over

MARKS | DO NOT WRITE IN THIS MARGIN

3. Two forces acting on a rocket are represented by vectors **u** and **v**.

$$\mathbf{u} = \begin{pmatrix} 2 \\ -5 \\ -3 \end{pmatrix} \text{ and } \mathbf{v} = \begin{pmatrix} 7 \\ 4 \\ -1 \end{pmatrix}.$$

Calculate $|\mathbf{u} + \mathbf{v}|$, the magnitude of the resultant force.

Express your answer as a surd in its simplest form.

3

MARKS | DO NOT WRITE IN THIS MARGIN

4. The diagram below shows part of the graph of $y = ax^2$.

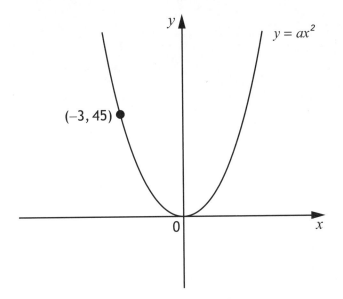

Find the value of a.

2

[Turn over

MARKS | DO NOT WRITE IN THIS MARGIN

5. Determine the nature of the roots of the function $f(x) = 7x^2 + 5x - 1$.

2

MARKS | DO NOT WRITE IN THIS MARGIN

6. A cattle farmer records the weight of some of his calves.

The scattergraph shows the relationship between the age, A months, and the weight, W kilograms, of the calves.

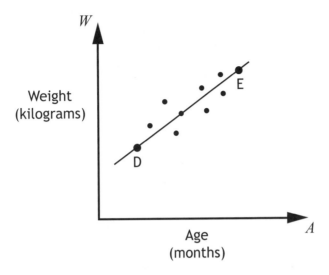

A line of best fit is drawn.

Point D represents a 3-month-old calf which weighs 100 kilograms.

Point E represents a 15-month-old calf which weighs 340 kilograms.

(a) Find the equation of the line of best fit in terms of A and W.

Give the equation in its simplest form.

3

[Turn over

MARKS | DO NOT WRITE IN THIS MARGIN

6. (continued)

(b) Use your equation from part (a) to estimate the weight of a 1-**year**-old calf.

Show your working. 1

MARKS | DO NOT WRITE IN THIS MARGIN

7. Ten couples took part in a dance competition.

    The couples were given a score in each round.

    The scores in the first round were

    16   27   12   18   26   21   27   22   18   17

    (a) Calculate the median and semi-interquartile range of these scores.   **3**

    (b) In the second round, the median was 26 and the semi-interquartile range was 2·5.

    Make two valid comparisons between the scores in the first and second rounds.   **2**

    [Turn over

MARKS | DO NOT WRITE IN THIS MARGIN

8. Two groups of people go to a theatre.

   Bill buys tickets for 5 adults and 3 children.

   The total cost of his tickets is £158·25.

   (a) Write down an equation to illustrate this information. **1**

   (b) Ben buys tickets for 3 adults and 2 children.

   The total cost of his tickets is £98.

   Write down an equation to illustrate this information. **1**

   (c) Calculate the cost of a ticket for an adult and the cost of a ticket for a child. **4**

MARKS | DO NOT WRITE IN THIS MARGIN

9. 480 000 tickets were sold for a tennis tournament last year.

    This represents 80% of all the available tickets.

    Calculate the total number of tickets that were available for this tournament.   **3**

10. The function $f(x)$ is defined by $f(x) = \dfrac{2}{\sqrt{x}}$, $x > 0$.

    Express $f(5)$ as a fraction with a rational denominator.   **2**

[**Turn over**

MARKS | DO NOT WRITE IN THIS MARGIN

11. In the diagram, OABCDE is a regular hexagon with centre M.

Vectors **a** and **b** are represented by $\overrightarrow{OA}$ and $\overrightarrow{OB}$ respectively.

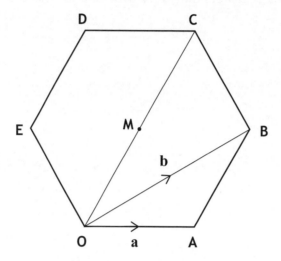

(a) Express $\overrightarrow{AB}$ in terms of **a** and **b**. 1

(b) Express $\overrightarrow{OC}$ in terms of **a** and **b**. 1

MARKS | DO NOT WRITE IN THIS MARGIN

12. Part of the graph of $y = a \sin bx°$ is shown in the diagram.

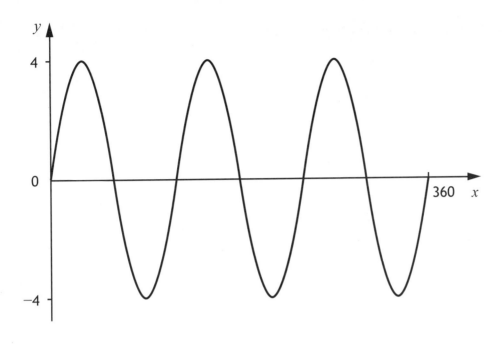

State the values of a and b.

2

[Turn over

MARKS | DO NOT WRITE IN THIS MARGIN

13. A parabola has equation $y = x^2 - 8x + 19$.

(a) Write the equation in the form $y = (x - p)^2 + q$.

2

(b) Sketch the graph of $y = x^2 - 8x + 19$, showing the coordinates of the turning point and the point of intersection with the y-axis.

3

14. Express

$$\frac{4}{x+2} - \frac{3}{x-4}, \qquad x \neq -2, \ x \neq 4$$

as a single fraction in its simplest form.

3

15. Simplify

$$\tan^2 x^\circ \cos^2 x^\circ.$$

Show your working.

2

[**Turn over**

MARKS | DO NOT WRITE IN THIS MARGIN

16. A cylindrical pipe has water in it as shown.

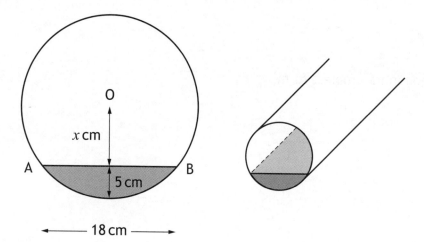

The depth of the water at the deepest point is 5 centimetres.

The width of the water surface, AB, is 18 centimetres.

The radius of the pipe is r centimetres.

The distance from the centre, O, of the pipe to the water surface is x centimetres.

(a) Write down an expression for x in terms of r.   **1**

(b) Calculate r, the radius of the pipe.   **3**

[END OF SPECIMEN QUESTION PAPER]

ADDITIONAL SPACE FOR ANSWERS

ADDITIONAL SPACE FOR ANSWERS

N5

National Qualifications
SPECIMEN ONLY

S847/75/02

Mathematics
Paper 2

Date — Not applicable

Duration — 1 hour 50 minutes

Fill in these boxes and read what is printed below.

Full name of centre

Town

Forename(s)

Surname

Number of seat

Date of birth

Day	Month	Year	Scottish candidate number

Total marks — 60

Attempt ALL questions.

You may use a calculator.

To earn full marks you must show your working in your answers.

State the units for your answer where appropriate.

Write your answers clearly in the spaces provided in this booklet. Additional space for answers is provided at the end of this booklet. If you use this space you must clearly identify the question number you are attempting.

Use **blue** or **black** ink.

Before leaving the examination room you must give this booklet to the Invigilator; if you do not, you may lose all the marks for this paper.

FORMULAE LIST

The roots of

$$ax^2 + bx + c = 0 \text{ are } x = \frac{-b \pm \sqrt{(b^2 - 4ac)}}{2a}$$

Sine rule:

$$\frac{a}{\sin A} = \frac{b}{\sin B} = \frac{c}{\sin C}$$

Cosine rule:

$$a^2 = b^2 + c^2 - 2bc\cos A \text{ or } \cos A = \frac{b^2 + c^2 - a^2}{2bc}$$

Area of a triangle:

$$A = \tfrac{1}{2}ab\sin C$$

Volume of a sphere:

$$V = \tfrac{4}{3}\pi r^3$$

Volume of a cone:

$$V = \tfrac{1}{3}\pi r^2 h$$

Volume of a pyramid:

$$V = \tfrac{1}{3}Ah$$

Standard deviation:

$$s = \sqrt{\frac{\Sigma(x - \overline{x})^2}{n-1}}$$

$$\text{or} \quad s = \sqrt{\frac{\Sigma x^2 - \frac{(\Sigma x)^2}{n}}{n-1}}, \text{ where } n \text{ is the sample size.}$$

MARKS | DO NOT WRITE IN THIS MARGIN

Total marks — 60

Attempt ALL questions

1. Beth normally cycles a total distance of 64 miles per week.

   She increases her total distance by 15% each week for the next three weeks.

   How many miles does she cycle in the third week?   **3**

   Give your answer to the nearest mile.

2. There are 3×10^5 platelets per millilitre of blood.

   On average, a person has 5·5 litres of blood.

   On average, how many platelets does a person have in their blood?   **2**

   Give your answer in scientific notation.

[Turn over

MARKS | DO NOT WRITE IN THIS MARGIN

3. Expand and simplify

$$(2x+3)(x^2-4x+1).$$

3

4. The diagram shows a cube placed on top of a cuboid, relative to the coordinate axes.

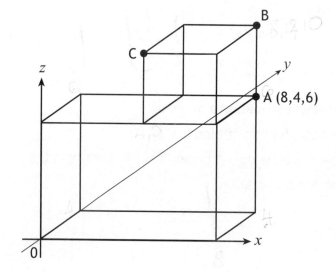

A is the point (8,4,6).

Write down the coordinates of B and C.

2

MARKS | DO NOT WRITE IN THIS MARGIN

5. In triangle PQR, PQ = 8 centimetres, QR = 3 centimetres and angle PQR = 120°.

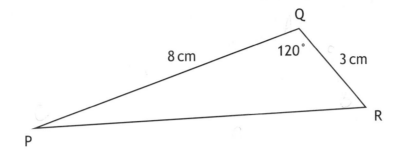

Calculate the length of PR.   3

[Turn over

MARKS | DO NOT WRITE IN THIS MARGIN

6. A child's toy is in the shape of a hemisphere with a cone on top, as shown in the diagram.

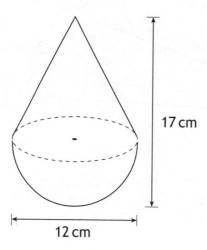

The toy is 12 centimetres wide and 17 centimetres high.

Calculate the volume of the toy.

Give your answer correct to 2 significant figures.

5

MARKS | DO NOT WRITE IN THIS MARGIN

7. Screenwash is available in bottles which are mathematically similar.

250 ml | 15 cm

36 cm

The smaller bottle has a height of 15 centimetres and a volume of 250 millilitres.

The larger bottle has a height of 36 centimetres.

Calculate the volume of the larger bottle.   3

[Turn over

8. Simplify $\dfrac{n^5 \times 10n}{2n^2}$.

3

MARKS | DO NOT WRITE IN THIS MARGIN

9. (a) A straight line has equation $4x + 3y = 12$.

Find the gradient of this line.

2

(b) State the coordinates of the point where the line crosses the y-axis.

1

[Turn over

MARKS | DO NOT WRITE IN THIS MARGIN

10. The top of a table is in the shape of a regular hexagon.

The three diagonals of the hexagon, which are shown as dotted lines in the diagram below, each have length 40 centimetres.

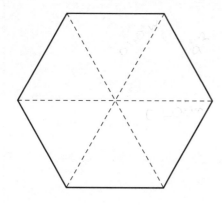 

Calculate the area of the top of the table.   4

MARKS | DO NOT WRITE IN THIS MARGIN

11. A cone is formed from a paper circle with a sector removed as shown.

The radius of the paper circle is 30 centimetres.

Angle AOB is 110°.

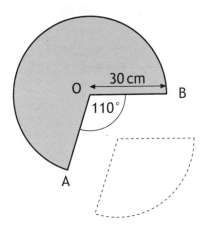

(a) Calculate the area of the sector removed from the circle.   3

(b) Calculate the circumference of the base of the cone.   3

[Turn over

MARKS | DO NOT WRITE IN THIS MARGIN

12. Part of the graph $y = 3\cos x° - 1$ is shown below.

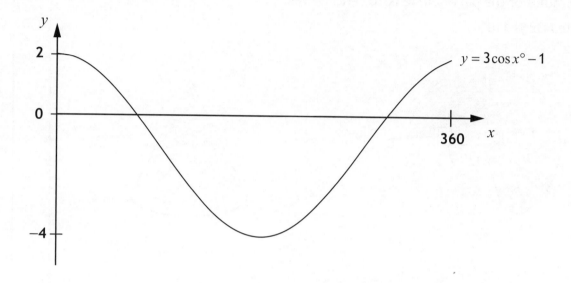

Calculate the x-coordinates of the points where the graph cuts the x-axis.

4

MARKS | DO NOT WRITE IN THIS MARGIN

13. Simplify $\dfrac{x^2 - 4x}{x^2 + x - 20}$.   3

[**Turn over**

14. Change the subject of the formula $s = ut + \dfrac{1}{2}at^2$ to a.

3

MARKS DO NOT WRITE IN THIS MARGIN

15. A yacht sails from a harbour H to a point C, then to a point D as shown below.

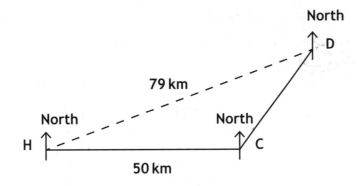

C is 50 kilometres due east of H.

D is on a bearing of 040° from C and is 79 kilometres from H.

(a) Calculate the size of angle CDH.   4

(b) Hence, calculate the bearing on which the yacht must sail to return directly to the harbour.   2

[Turn over

Page fifteen

MARKS | DO NOT WRITE IN THIS MARGIN

16. A rectangular picture measuring 9 centimetres by 13 centimetres is placed on a rectangular piece of card.

The area of the card is 270 square centimetres.

There is a border x centimetres wide on all sides of the picture.

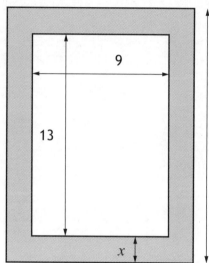

(a) (i) Write down an expression for the length of the card in terms of x.   1

(ii) Hence show that $4x^2 + 44x - 153 = 0$.   2

MARKS | DO NOT WRITE IN THIS MARGIN

16. **(continued)**

(b) Calculate x, the width of the border.

Give your answer correct to one decimal place.   4

[END OF SPECIMEN QUESTION PAPER]

MARKS DO NOT WRITE IN THIS MARGIN

ADDITIONAL SPACE FOR ANSWERS

NATIONAL 5

2018

N5

National Qualifications 2018

Mark

X847/75/01

**Mathematics
Paper 1
(Non-Calculator)**

FRIDAY, 4 MAY

9:00 AM – 10:15 AM

Fill in these boxes and read what is printed below.

Full name of centre

Town

Forename(s)

Surname

Number of seat

Date of birth

Day	Month	Year	Scottish candidate number

Total marks — 50

Attempt ALL questions.

You may NOT use a calculator.

To earn full marks you must show your working in your answers.

State the units for your answer where appropriate.

Write your answers clearly in the spaces provided in this booklet. Additional space for answers is provided at the end of this booklet. If you use this space you must clearly identify the question number you are attempting.

Use **blue** or **black** ink.

Before leaving the examination room you must give this booklet to the Invigilator; if you do not, you may lose all the marks for this paper.

FORMULAE LIST

The roots of $ax^2 + bx + c = 0$ are $x = \dfrac{-b \pm \sqrt{(b^2 - 4ac)}}{2a}$

Sine rule: $\dfrac{a}{\sin A} = \dfrac{b}{\sin B} = \dfrac{c}{\sin C}$

Cosine rule: $a^2 = b^2 + c^2 - 2bc\cos A$ or $\cos A = \dfrac{b^2 + c^2 - a^2}{2bc}$

Area of a triangle: $A = \frac{1}{2}ab\sin C$

Volume of a sphere: $V = \frac{4}{3}\pi r^3$

Volume of a cone: $V = \frac{1}{3}\pi r^2 h$

Volume of a pyramid: $V = \frac{1}{3}Ah$

Standard deviation: $s = \sqrt{\dfrac{\Sigma(x - \bar{x})^2}{n-1}}$

or $s = \sqrt{\dfrac{\Sigma x^2 - \dfrac{(\Sigma x)^2}{n}}{n-1}}$, where n is the sample size.

MARKS | DO NOT WRITE IN THIS MARGIN

Total marks — 50

Attempt ALL questions

1. Evaluate $2\frac{1}{3} + \frac{4}{5}$.

   2

2. Expand and simplify $(3x+1)(x-1) + 2(x^2 - 5)$.

   3

[Turn over

MARKS | DO NOT WRITE IN THIS MARGIN

3. Solve, algebraically, the system of equations

$$4x + 5y = -3$$
$$6x - 2y = 5.$$

3

4. Two vectors are given by $\mathbf{u} = \begin{pmatrix} 1 \\ 5 \\ 1 \end{pmatrix}$ and $\mathbf{u} + \mathbf{v} = \begin{pmatrix} 6 \\ -4 \\ 3 \end{pmatrix}$.

Find vector $\mathbf{v}$.

Express your answer in component form.

2

MARKS | DO NOT WRITE IN THIS MARGIN

5. Solve

$$x^2 - 11x + 24 = 0.$$

2

6. Part of the graph of $y = a\cos bx°$ is shown in the diagram.

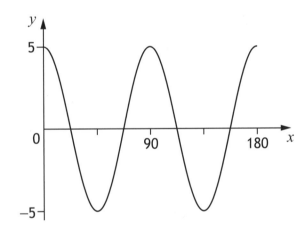

State the values of a and b.

2

[Turn over

MARKS | DO NOT WRITE IN THIS MARGIN

7. The cost of a journey with Tom's Taxis depends on the distance travelled.

The graph below shows the cost, P pounds, of a journey with Tom's Taxis against the distance travelled, d miles.

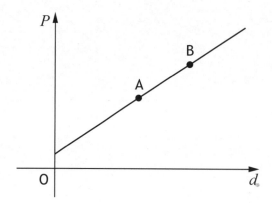

Point A represents a journey of 8 miles which costs £14.
Point B represents a journey of 12 miles which costs £20.

(a) Find the equation of the line in terms of P and d.

Give the equation in its simplest form.

3

MARKS | DO NOT WRITE IN THIS MARGIN

7. (continued)

(b) Calculate the cost of a journey of 5 miles.   **1**

8. Determine the nature of the roots of the function $f(x) = 2x^2 + 4x + 5$.   **2**

[Turn over

MARKS | DO NOT WRITE IN THIS MARGIN

9. In the diagram shown below, ABCDEFGHJK is a regular decagon.
    • Angle KLJ is 17°.
    • AKL is a straight line.

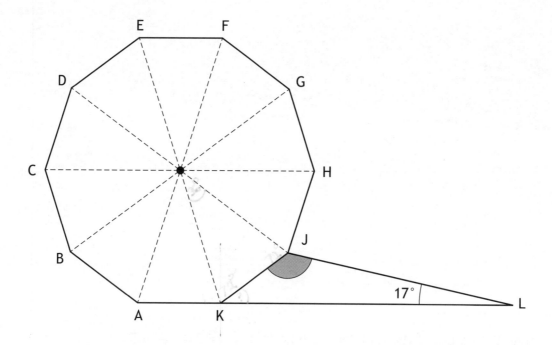

Calculate the size of shaded angle KJL.

2

MARKS | DO NOT WRITE IN THIS MARGIN

10. In triangle XYZ:

- XZ = 10 centimetres
- YZ = 8 centimetres
- $\cos Z = \dfrac{1}{8}$.

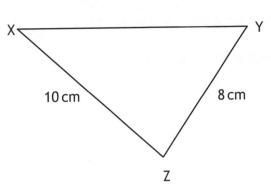

Calculate the length of XY.

3

[Turn over

MARKS | DO NOT WRITE IN THIS MARGIN

11. Express $\dfrac{9}{\sqrt{6}}$ with a rational denominator.

Give your answer in its simplest form.

2

12. Given that $\cos 60° = 0{\cdot}5$, state the value of $\cos 240°$.

1

MARKS | DO NOT WRITE IN THIS MARGIN

13. The diagram shows a triangular prism, ABCDEF, relative to the coordinate axes.

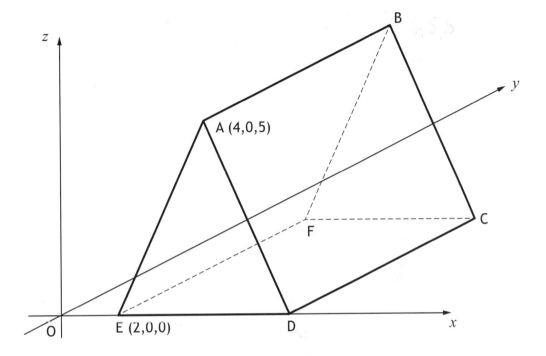

- AD = AE.
- DC = 8 units.
- Edges EF, DC and AB are parallel to the y-axis.

Write down the coordinates of B and C.
    2

[Turn over

MARKS | DO NOT WRITE IN THIS MARGIN

14. Change the subject of the formula $y = g\sqrt{x} + h$ to x.

3

15. Remove the brackets and simplify $\left(\dfrac{2}{3}p^4\right)^2$.

2

MARKS | DO NOT WRITE IN THIS MARGIN

16. Sketch the graph of $y = (x-6)(x+4)$.

On your sketch, show clearly the points of intersection with the x-axis and the y-axis, and the coordinates of the turning point. **3**

[Turn over

MARKS | DO NOT WRITE IN THIS MARGIN

17. A square based pyramid is shown in the diagram below.

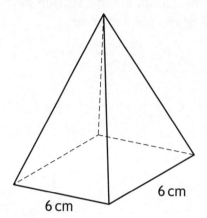

6 cm

6 cm

The square base has length 6 centimetres.
The volume is 138 cubic centimetres.
Calculate the height of the pyramid.

3

MARKS | DO NOT WRITE IN THIS MARGIN

18. Express $\sin x^\circ \cos x^\circ \tan x^\circ$ in its simplest form.

Show your working.

2

[Turn over

MARKS | DO NOT WRITE IN THIS MARGIN

19. (a) (i) Express $x^2 - 6x - 81$ in the form $(x - p)^2 + q$.

2

(ii) Hence state the equation of the axis of symmetry of the graph of $y = x^2 - 6x - 81$.

1

MARKS | DO NOT WRITE IN THIS MARGIN

19. (continued)

(b) The roots of the equation $x^2 - 6x - 81 = 0$ can be expressed in the form $x = d \pm d\sqrt{e}$.

Find, algebraically, the values of d and e.

4

[END OF QUESTION PAPER]

ADDITIONAL SPACE FOR ANSWERS

MARKS | DO NOT WRITE IN THIS MARGIN

ADDITIONAL SPACE FOR ANSWERS

[BLANK PAGE]

DO NOT WRITE ON THIS PAGE

National Qualifications 2018

Mark

X847/75/02

Mathematics Paper 2

FRIDAY, 4 MAY

10:35 AM — 12:25 PM

Fill in these boxes and read what is printed below.

Full name of centre

Town

Forename(s)

Surname

Number of seat

Date of birth

Day	Month	Year	Scottish candidate number

Total marks — 60

Attempt ALL questions.

You may use a calculator.

To earn full marks you must show your working in your answers.

State the units for your answer where appropriate.

Write your answers clearly in the spaces provided in this booklet. Additional space for answers is provided at the end of this booklet. If you use this space you must clearly identify the question number you are attempting.

Use **blue** or **black** ink.

Before leaving the examination room you must give this booklet to the Invigilator; if you do not, you may lose all the marks for this paper.

FORMULAE LIST

The roots of $ax^2+bx+c=0$ are $x=\dfrac{-b\pm\sqrt{(b^2-4ac)}}{2a}$

Sine rule: $\dfrac{a}{\sin A}=\dfrac{b}{\sin B}=\dfrac{c}{\sin C}$

Cosine rule: $a^2=b^2+c^2-2bc\cos A$ or $\cos A=\dfrac{b^2+c^2-a^2}{2bc}$

Area of a triangle: $A=\frac{1}{2}ab\sin C$

Volume of a sphere: $V=\frac{4}{3}\pi r^3$

Volume of a cone: $V=\frac{1}{3}\pi r^2 h$

Volume of a pyramid: $V=\frac{1}{3}Ah$

Standard deviation: $s=\sqrt{\dfrac{\Sigma(x-\bar{x})^2}{n-1}}$

or $s=\sqrt{\dfrac{\Sigma x^2-\dfrac{(\Sigma x)^2}{n}}{n-1}}$, where n is the sample size.

MARKS | DO NOT WRITE IN THIS MARGIN

Total marks — 60

Attempt ALL questions

1. Households in a city produced a total of 125 000 tonnes of waste in 2017.

   The total amount of waste is expected to fall by 2% each year.

   Calculate the total amount of waste these households are expected to produce in 2020.

   3

[Turn over

MARKS | DO NOT WRITE IN THIS MARGIN

2. The diagram below shows a sector of a circle, centre C.

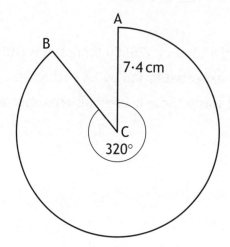

The radius of the circle is 7·4 centimetres.

Calculate the length of the major arc AB.

3

MARKS | DO NOT WRITE IN THIS MARGIN

3. Find $|\mathbf{r}|$, the magnitude of vector $\mathbf{r} = \begin{pmatrix} 24 \\ -12 \\ 8 \end{pmatrix}$.

2

4. Solve, algebraically, the inequation

$$3x < 6(x-1) - 12.$$

3

[Turn over

MARKS | DO NOT WRITE IN THIS MARGIN

5. A farmers' market took place one weekend.

Stallholders were asked to record the number of customers who visited their stall.

The number of customers who visited six of the stalls on Saturday were as follows:

120   126   125   131   130   124

(a) Calculate the mean and standard deviation of the number of customers.   **4**

MARKS | DO NOT WRITE IN THIS MARGIN

5. (continued)

The mean number of customers who visited these six stalls on Sunday was 117 and the standard deviation was 6·2.

(b) Make two valid comments comparing the number of customers who visited these stalls on Saturday and Sunday.

2

6. A function is defined as $f(x) = 5 + 4x$.

Given that $f(a) = 73$, calculate a.

2

[Turn over

MARKS | DO NOT WRITE IN THIS MARGIN

7. A toy company makes juggling balls in the shape of a sphere with a diameter of 6·4 centimetres.

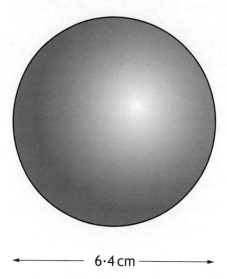

◄────── 6·4 cm ──────►

Calculate the volume of one juggling ball.

Give your answer correct to 2 significant figures.

3

MARKS | DO NOT WRITE IN THIS MARGIN

8. Solve the equation $7\sin x° + 2 = 3$, for $0 \le x < 360$.

3

[Turn over

MARKS | DO NOT WRITE IN THIS MARGIN

9. In this diagram:

    • angle ABD = 75°

    • angle BDC = 37°

    • BC = 20 centimetres.

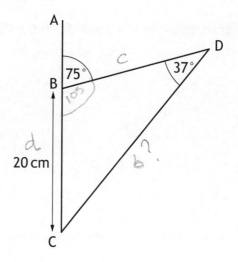

Calculate the length of DC.   3

$$\frac{b}{\sin 105} = \frac{20}{\sin 37} \quad \therefore \frac{b \times \sin 37}{\sin 37} = \frac{20 \times \sin 105}{\sin 37}$$

$$= b = \frac{20 \times \sin 105}{\sin 37}$$

$$= 32.1 \text{ cm}.$$

MARKS | DO NOT WRITE IN THIS MARGIN

10. In the diagram below, $\overrightarrow{AB}$ and $\overrightarrow{EA}$ represent the vectors **u** and **w** respectively.

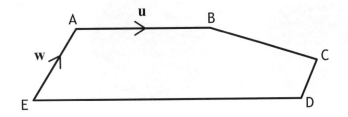

- $\overrightarrow{ED} = 2\overrightarrow{AB}$
- $\overrightarrow{EA} = 2\overrightarrow{DC}$

Express $\overrightarrow{BC}$ in terms of **u** and **w**.

Give your answer in its simplest form.   2

[Turn over

MARKS | DO NOT WRITE IN THIS MARGIN

11. Venus and Earth are two planets within our solar system.

Venus   Earth

The volume of Venus is approximately $9 \cdot 3 \times 10^{11}$ cubic kilometres.

This is 85% of the volume of Earth.

Calculate the volume of Earth.

3

MARKS | DO NOT WRITE IN THIS MARGIN

12. The shape below is part of a circle, centre O.

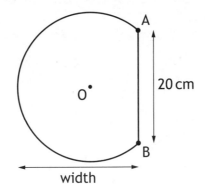

20 cm

width

The circle has radius 13 centimetres.

AB is a chord of length 20 centimetres.

Calculate the width of the shape.

4

[Turn over

MARKS | DO NOT WRITE IN THIS MARGIN

13. A ferry and a trawler receive a request for help from a stranded yacht.

On the diagram the points F, T and Y show the positions of the ferry, the trawler and the yacht respectively.

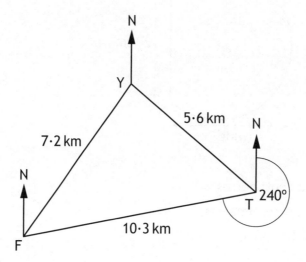

- FY is 7·2 kilometres.

- TY is 5·6 kilometres.

- FT is 10·3 kilometres.

- F is on a bearing of 240° from T.

Calculate the bearing of the yacht from the trawler.   4

MARKS | DO NOT WRITE IN THIS MARGIN

14. A straight line has equation $2x - 5y = 20$.

Find the coordinates of the point where this line crosses the y-axis.   **2**

15. Express

$$\frac{n}{n^2 - 4} \div \frac{3}{n-2}, \qquad n \neq -2,\ n \neq 2$$

as a single fraction in its simplest form.   **3**

[Turn over

MARKS | DO NOT WRITE IN THIS MARGIN

16. Chris wants to store his umbrella in a locker.

The locker is a cuboid with internal dimensions of length 40 centimetres, breadth 40 centimetres and height 70 centimetres.

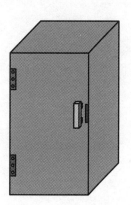
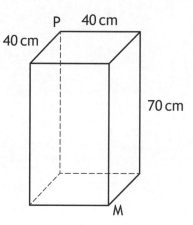

The umbrella is 85 centimetres long.

He thinks it will fit into the locker from corner P to corner M.

Is he correct?

Justify your answer.   4

MARKS | DO NOT WRITE IN THIS MARGIN

17. In the diagram below AOD is a sector of a circle, with centre O, and BOC is a triangle.

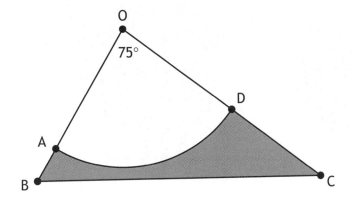

In sector AOD:

- radius = 30 centimetres
- angle AOD = 75°.

In triangle OBC:

- OB = 38 centimetres
- OC = 55 centimetres.

Calculate the area of the shaded region, ABCD.   5

[Turn over

18. A cinema sells popcorn in two different sized cartons.

16 cm

24 cm

The small carton is 16 centimetres deep and has a volume of 576 cubic centimetres.

The large carton is 24 centimetres deep and has a volume of 1125 cubic centimetres.

(a) Show that the two cartons are **not** mathematically similar.　　　3

MARKS | DO NOT WRITE IN THIS MARGIN

18. **(continued)**

The large carton is redesigned so that the two cartons are **now** mathematically similar.

The volume of the redesigned large carton is 1500 cubic centimetres.

(b) Calculate the depth of the redesigned large carton.   2

[END OF QUESTION PAPER]

ADDITIONAL SPACE FOR ANSWERS

MARKS | DO NOT WRITE IN THIS MARGIN

ADDITIONAL SPACE FOR ANSWERS

[BLANK PAGE]

DO NOT WRITE ON THIS PAGE

NATIONAL 5

2019

National
Qualifications
2019

Mark

X847/75/01

Mathematics
Paper 1
(Non-Calculator)

FRIDAY, 3 MAY

9:00 AM — 10:15 AM

Fill in these boxes and read what is printed below.

Full name of centre

Town

Forename(s)

Surname

Number of seat

Date of birth

Day   Month   Year   Scottish candidate number

Total marks — 50

Attempt ALL questions.

You may NOT use a calculator.

To earn full marks you must show your working in your answers.

State the units for your answer where appropriate.

Write your answers clearly in the spaces provided in this booklet. Additional space for answers is provided at the end of this booklet. If you use this space you must clearly identify the question number you are attempting.

Use **blue** or **black** ink.

Before leaving the examination room you must give this booklet to the Invigilator; if you do not, you may lose all the marks for this paper.

FORMULAE LIST

The roots of $ax^2 + bx + c = 0$ are $x = \dfrac{-b \pm \sqrt{(b^2 - 4ac)}}{2a}$

Sine rule: $\dfrac{a}{\sin A} = \dfrac{b}{\sin B} = \dfrac{c}{\sin C}$

Cosine rule: $a^2 = b^2 + c^2 - 2bc\cos A$ or $\cos A = \dfrac{b^2 + c^2 - a^2}{2bc}$

Area of a triangle: $A = \tfrac{1}{2} ab \sin C$

Volume of a sphere: $V = \tfrac{4}{3}\pi r^3$

Volume of a cone: $V = \tfrac{1}{3}\pi r^2 h$

Volume of a pyramid: $V = \tfrac{1}{3} Ah$

Standard deviation: $s = \sqrt{\dfrac{\Sigma(x - \bar{x})^2}{n-1}}$

or $\quad s = \sqrt{\dfrac{\Sigma x^2 - \dfrac{(\Sigma x)^2}{n}}{n-1}}$, where n is the sample size.

MARKS

Total marks — 50

Attempt ALL questions

1. Given that $f(x) = 5x^3$, evaluate $f(-2)$.

2

2. Evaluate $\dfrac{3}{8} \times 1\dfrac{5}{7}$.

Give your answer in its simplest form.

2

3. Expand and simplify $(x+5)(2x^2-7x-3)$.

3

4. The diagram below shows a sector of a circle, centre C.

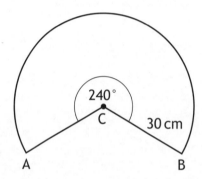

The radius of the circle is 30 centimetres.

Calculate the length of the major arc AB.

Take $\pi = 3\cdot14$.

3

MARKS DO NOT WRITE IN THIS MARGIN

5. The midday temperatures in Grantford were recorded over a nine day period.

The temperatures, in °C, were

4   7   4   3   6   10   9   5   3

  (a) Calculate the median and semi-interquartile range for these temperatures.   **3**

Over the same nine day period the midday temperatures in Endoch were also recorded.

The median temperature was 8 °C, and the semi-interquartile range was 1·5 °C.

  (b) Make two valid comments comparing the midday temperatures of Grantford and Endoch during this period.   **2**

MARKS | DO NOT WRITE IN THIS MARGIN

6. The fuel consumption of a group of cars is recorded.

    The scattergraph shows the relationship between the fuel consumption, F kilometres per litre, and the engine size, E litres, of the cars.

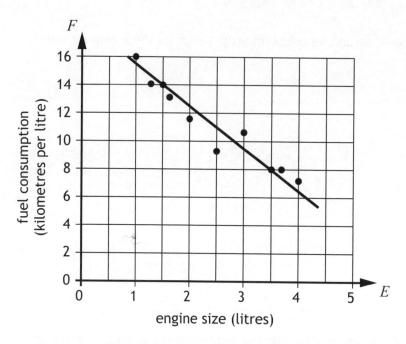

    engine size (litres)

    A line of best fit has been drawn.

    (a) Find the equation of the line of best fit in terms of F and E.

         Give the equation in its simplest form.   3

MARKS | DO NOT WRITE IN THIS MARGIN

6. **(continued)**

Amaar's car has an engine size of 1·1 litres.

(b) Use your equation from part (a) to estimate how many kilometres per litre he should expect to get.

1

7. The area of a trapezium is given by the formula

$$A = \frac{1}{2}h(x+y).$$

Make x the subject of the formula.

3

MARKS | DO NOT WRITE IN THIS MARGIN

8. John bought 7 bags of cement and 3 bags of gravel.

The total weight of these bags was 215 kilograms.

(a) Write down an equation to illustrate this information.   **1**

Shona bought 5 bags of cement and 4 bags of gravel.

The total weight of her bags was 200 kilograms.

(b) Write down an equation to illustrate this information.   **1**

(c) Calculate the weight of one bag of cement and the weight of one bag of gravel.   **4**

MARKS | DO NOT WRITE IN THIS MARGIN

9. The graph shows a parabola.

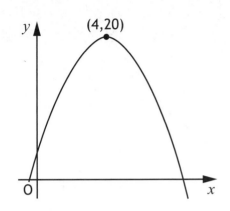

The maximum turning point has coordinates (4,20) as shown in the diagram.

(a) Write down the equation of the axis of symmetry of the graph.   **1**

The equation of the parabola is of the form $y = b - (x + a)^2$.

(b) State the values of

(i) a **1**

(ii) b.   **1**

MARKS | DO NOT WRITE IN THIS MARGIN

10. In triangle PQR, $\overrightarrow{PR} = \begin{pmatrix} 6 \\ -4 \end{pmatrix}$ and $\overrightarrow{RQ} = \begin{pmatrix} -1 \\ 8 \end{pmatrix}$.

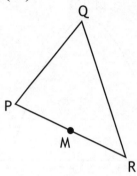

(a) Express $\overrightarrow{PQ}$ in component form.

1

M is the midpoint of PR.

(b) Express $\overrightarrow{MQ}$ in component form.

2

MARKS | DO NOT WRITE IN THIS MARGIN

11. Pam is designing a company logo.

She starts by drawing a regular pentagon ABCDE.

The vertices of the pentagon lie on the circumference of a circle with centre O.

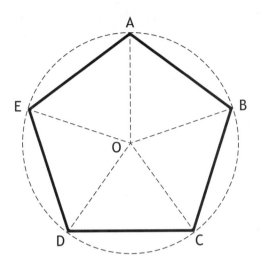

She then adds to the design as shown in the diagram below.

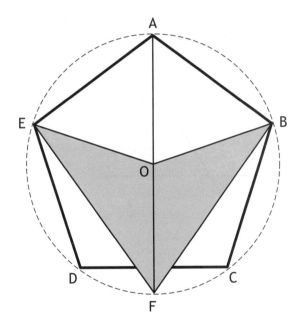

AF is a diameter of the circle.

Calculate the size of angle OFB.

3

MARKS | DO NOT WRITE IN THIS MARGIN

12. Express $\dfrac{\sqrt{2}}{\sqrt{40}}$ as a fraction with a rational denominator.

Give your answer in its simplest form.

3

13. Part of the graph of $y = 3\cos(x + 45)°$ is shown in the diagram.

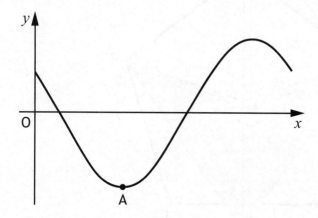

The graph has a minimum turning point at A.

State the coordinates of A.

2

MARKS | DO NOT WRITE IN THIS MARGIN

14. Solve the equation $\dfrac{x}{2} - 1 = \dfrac{3-x}{5}$.

3

MARKS | DO NOT WRITE IN THIS MARGIN

15. A ball is kicked from a clifftop.

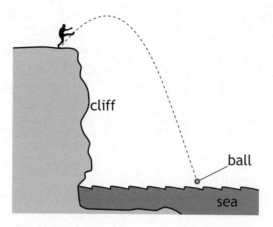

The height, h metres, of the ball relative to the clifftop after t seconds is given by $h = 12t - 5t^2$.

(a) Calculate the height of the ball above the clifftop after 2 seconds.

1

MARKS | DO NOT WRITE IN THIS MARGIN

15. (continued)

The graph below represents the height, h metres, of the ball relative to the clifftop after t seconds.

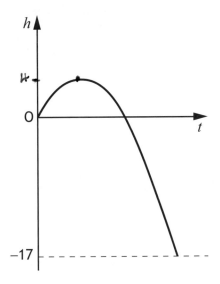

The sea is 17 metres below the clifftop.

(b) After how many seconds will the ball hit the sea?

4

[END OF QUESTION PAPER]

ADDITIONAL SPACE FOR ANSWERS

MARKS | DO NOT WRITE IN THIS MARGIN

ADDITIONAL SPACE FOR ANSWERS

[BLANK PAGE]

DO NOT WRITE ON THIS PAGE

[BLANK PAGE]

DO NOT WRITE ON THIS PAGE

[BLANK PAGE]

DO NOT WRITE ON THIS PAGE

N5

National Qualifications 2019

Mark

X847/75/02

Mathematics Paper 2

FRIDAY, 3 MAY

10:45 AM — 12:35 PM

Fill in these boxes and read what is printed below.

Full name of centre

Town

Forename(s)

Surname

Number of seat

Date of birth

Day	Month	Year	Scottish candidate number

Total marks — 60

Attempt ALL questions.

You may use a calculator.

To earn full marks you must show your working in your answers.

State the units for your answer where appropriate.

Write your answers clearly in the spaces provided in this booklet. Additional space for answers is provided at the end of this booklet. If you use this space you must clearly identify the question number you are attempting.

Use **blue** or **black** ink.

Before leaving the examination room you must give this booklet to the Invigilator; if you do not, you may lose all the marks for this paper.

FORMULAE LIST

The roots of $ax^2 + bx + c = 0$ are $x = \dfrac{-b \pm \sqrt{(b^2 - 4ac)}}{2a}$

Sine rule: $\dfrac{a}{\sin A} = \dfrac{b}{\sin B} = \dfrac{c}{\sin C}$

Cosine rule: $a^2 = b^2 + c^2 - 2bc\cos A$ or $\cos A = \dfrac{b^2 + c^2 - a^2}{2bc}$

Area of a triangle: $A = \tfrac{1}{2}ab\sin C$

Volume of a sphere: $V = \tfrac{4}{3}\pi r^3$

Volume of a cone: $V = \tfrac{1}{3}\pi r^2 h$

Volume of a pyramid: $V = \tfrac{1}{3}Ah$

Standard deviation: $s = \sqrt{\dfrac{\Sigma(x - \bar{x})^2}{n-1}}$

or $\quad s = \sqrt{\dfrac{\Sigma x^2 - \dfrac{(\Sigma x)^2}{n}}{n-1}}$, where n is the sample size.

MARKS | DO NOT WRITE IN THIS MARGIN

Total marks — 60

Attempt ALL questions

1. A charity distributed 80 000 emergency packages during 2018.

   This number is expected to increase by 15% each year.

   Calculate how many emergency packages the charity expects to distribute in 2021.   **3**

2. Find $|\mathbf{p}|$, the magnitude of vector $\mathbf{p} = \begin{pmatrix} 6 \\ 27 \\ -18 \end{pmatrix}$.   **2**

MARKS | DO NOT WRITE IN THIS MARGIN

3. The diagram shows triangle PQR.

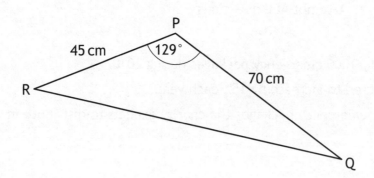

- PR = 45 centimetres
- PQ = 70 centimetres
- Angle QPR = 129°

Calculate the area of triangle PQR.

2

4. A sesame seed weighs $3{\cdot}6 \times 10^{-6}$ kilograms.

The weight of a poppy seed is 8% of the weight of a sesame seed.

Calculate the weight of a poppy seed in kilograms.

Give your answer in scientific notation.

2

MARKS | DO NOT WRITE IN THIS MARGIN

5. The diagram shows a cone with diameter 6 units and height 8 units.

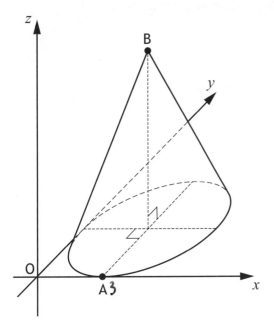

- The x-axis and the y-axis are tangents to the base
- A is the point of contact between the base and the x-axis
- B is directly above the centre of the base

Write down the coordinates of A and B.   2

6. Solve the equation $3x^2 + 9x - 2 = 0$.

   Give your answers correct to 1 decimal place.

   3

MARKS | DO NOT WRITE IN THIS MARGIN

7. Triangle XYZ is shown below.

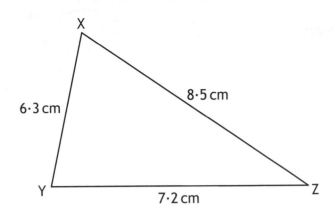

Calculate the size of the smallest angle in triangle XYZ.

3

MARKS | DO NOT WRITE IN THIS MARGIN

8. A traffic bollard is in the shape of a cylinder with a hemisphere on top.

   The bollard has

   - diameter 24 centimetres
   - height 70 centimetres.

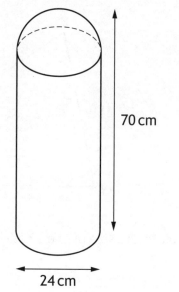

70 cm

24 cm

Calculate the volume of the bollard.

Give your answer correct to 3 significant figures.

5

MARKS | DO NOT WRITE IN THIS MARGIN

9. Georgie had her roof repaired.

    She was charged an extra 2·5% for late payment.

    She had to pay a total of £977·85.

    Calculate how much she would have **saved** if she had paid on time.   3

10. Express $x^2 + 10x - 15$ in the form $(x + p)^2 + q$.   2

MARKS | DO NOT WRITE IN THIS MARGIN

11. The diagram shows the course for a jet-ski race.

The course is indicated by markers A, B and C.

The total length of the course is 1500 metres.

- B is 600 metres from A
- C is 650 metres from A
- C is due north of B

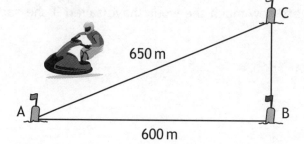

Determine whether B is due east of A.

Justify your answer.   4

MARKS | DO NOT WRITE IN THIS MARGIN

12. In the diagram
- ABC is a sector of a circle, centre C
- DEF is a sector of a circle, centre F.

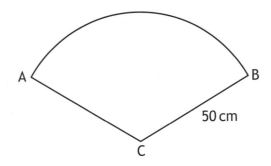 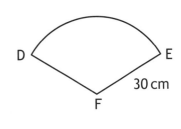

The sectors are mathematically similar.

The area of the larger sector, ABC, is 2750 square centimetres.

(a) Calculate the area of the smaller sector, DEF.   3

(b) Calculate the size of angle ACB.   3

MARKS | DO NOT WRITE IN THIS MARGIN

13. Find an expression for the gradient of the line joining point A(6,9) to point B($4p, 4p^2$).

Give your answer in its simplest form.   **3**

14. Solve the equation $5\cos x° + 2 = 1, \quad 0 \le x < 360$.   **3**

MARKS | DO NOT WRITE IN THIS MARGIN

15. Express

$$\frac{4}{x-2} - \frac{3}{x+5}, \quad x \neq 2, x \neq -5$$

as a single fraction in its simplest form.

3

16. Simplify $\dfrac{a^4 \times 3a}{\sqrt{a}}$.

3

[Turn over

MARKS | DO NOT WRITE IN THIS MARGIN

17. Expand and simplify

$$\left(\sin x^\circ + \cos x^\circ\right)^2.$$

Show your working.

2

18. The picture shows a cartoon snowman.

The diagram below represents the snowman.

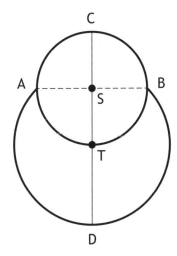

- The head is a small circle, centre S, with diameter 15 centimetres
- The body is part of a larger circle, centre T
- The point T lies on the circumference of the small circle
- The points A and B lie on the circumferences of both circles

Calculate CD, the height of the snowman.   4

MARKS | DO NOT WRITE IN THIS MARGIN

19. Katy and Mona are looking up at a hot-air balloon.

In the diagram below, K, M and B represent the positions of Katy, Mona and the balloon respectively.

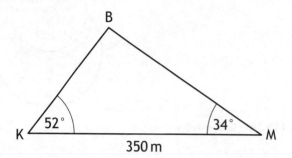

- The angle of elevation of the balloon from Katy is 52°
- The angle of elevation of the balloon from Mona is 34°
- Katy and Mona are 350 metres apart on level ground

Calculate the height of the hot-air balloon above the ground.

5

[END OF QUESTION PAPER]

MARKS | DO NOT WRITE IN THIS MARGIN

ADDITIONAL SPACE FOR ANSWERS

MARKS DO NOT WRITE IN THIS MARGIN

ADDITIONAL SPACE FOR ANSWERS

[BLANK PAGE]

DO NOT WRITE ON THIS PAGE

[BLANK PAGE]

DO NOT WRITE ON THIS PAGE

NATIONAL 5

Answers

SQA NATIONAL 5 MATHEMATICS 2019

NATIONAL 5 MATHEMATICS 2017 SPECIMEN QUESTION PAPER

One mark is available for each •. There are no half marks.

Paper 1

1. • $\dfrac{19}{8} \times \dfrac{16}{5}$

   • $7\dfrac{3}{5}$ or $\dfrac{38}{5}$

2. • $11 - 2 - 6x < 39$

   • $-6x < 30$ or $-30 < 6x$

   • $x > -5$ or $-5 < x$

3. • $\begin{pmatrix} 9 \\ -1 \\ -4 \end{pmatrix}$

   • $\sqrt{9^2 + (-1)^2 + (-4)^2}$

   • $7\sqrt{2}$

4. • $45 = a(-3)^2$ or equivalent

   • $a = 5$

5. • $53 \quad [5^2 - 4 \times 7 \times (-1)]$

   • two real and distinct roots

6. (a) • $\dfrac{240}{12}$ or equivalent

   • $y - 100 = \dfrac{240}{12}(x - 3)$

   or $y - 340 = \dfrac{240}{12}(x - 15)$

   or $100 = \dfrac{240}{12} \times 3 + c$

   or $340 = \dfrac{240}{12} \times 15 + c$

   • $W = 20A + 40$ or equivalent

   (b) • $20 \times 12 + 40 = 280\,\text{kg}$

7. (a) • median = 19·5

   • quartiles = 17 and 26

   • SIQR = 4·5

   (b) • On average the second round's scores are higher

   • The second round's scores are more consistent.

8. (a) • $5a + 3c = 158 \cdot 25$

   (b) • $3a + 2c = 98$

   (c) • e.g. $\begin{aligned} 10a + 6c &= 316 \cdot 50 \\ 9a + 6c &= 294 \end{aligned}$

   • values for a and c

   • $a = 22 \cdot 5$ and $c = 15 \cdot 25$

   • Adult £22·50
     Child £15·25

9. • $80\% = 480000$

   • $10\% = 60000$ or equivalent

   • 600000

10. • $\dfrac{2}{\sqrt{5}}$

    • $\dfrac{2\sqrt{5}}{5}$

11. (a) • $\mathbf{b} - \mathbf{a}$ or $-\mathbf{a} + \mathbf{b}$

    (b) • $2(\mathbf{b} - \mathbf{a})$ or $2(-\mathbf{a} + \mathbf{b})$

12. • $a = 4$

    • $b = 3$

13. (a) • $(x - 4)^2 \ldots\ldots\ldots$

    • $(x - 4)^2 + 3$

    (b)

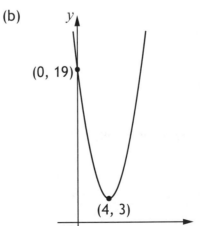

14. • $(x+2)(x-4)$

• $4(x-4)-3(x+2)$

• $\dfrac{x-22}{(x+2)(x-4)}$

15. • $\dfrac{\sin x}{\cos x}$ or $\dfrac{\sin^2 x}{\cos^2 x}$

• $\dfrac{\sin^2 x}{\cos^2 x} \times \cos^2 x = \sin^2 x$

16. (a) • $r-5$

(b) • $r^2 = (r-5)^2 + 9^2$

• $r^2 = r^2 - 10r + 25 + 81$

• $r = 10 \cdot 6$

Paper 2

1. • $\times 1 \cdot 15$

• $64 \times 1 \cdot 15^3$

• 97 miles

2. • $3 \times 10^5 \times 5 \cdot 5 \times 1000$

• $1 \cdot 65 \times 10^9$

3. • e.g. $2x^3 - 8x^2 + 2x \ldots$

• e.g. $\ldots 3x^2 - 12x + 3$

• $2x^3 - 5x^2 - 10x + 3$

4. • $B(8, 4, 10)$

• $C(4, 0, 10)$

5. • $(PR^2 =) 8^2 + 3^2 - 2 \times 8 \times 3 \times \cos 120°$

• 97

• $9 \cdot 8 \ (488\ldots)$ cm

6. • $\dfrac{1}{3} \times \pi \times 6^2 \times 11 (= 414 \cdot 690\ldots)$

• $\dfrac{4}{3} \times \pi \times 6^3 (= 904 \cdot 778\ldots)$ **or**

$\dfrac{1}{2} \times \dfrac{4}{3} \times \pi \times 6^3 (= 452 \cdot 389\ldots)$

• evidence

• $867 \cdot 079\ldots$

• 870 cm^3

7. • $\dfrac{36}{15} \ (= 2 \cdot 4)$

• $\left(\dfrac{36}{15}\right)^3 \times 250 \ (= 2 \cdot 4^3 \times 250)$

• 3456 ml

8. • $10n^6$

• $\dfrac{5n^6}{n^2}$

• $5n^4$

9. (a) • $3y = -4x + 12$

• gradient $= -\dfrac{4}{3}$

(b) • $(0, 4)$

10. • 60

• $\dfrac{1}{2} \times 20 \times 20 \times \sin 60$

• $\left(\dfrac{1}{2} \times 20 \times 20 \times \sin 60\right) \times 6$

• 1039$\cdot$2 cm^2

11. (a) • $\dfrac{110}{360}$

• $\dfrac{110}{360} \times \pi \times 30^2$

• $863 \cdot 9\ldots$ / 864 cm^2

(b) • $\dfrac{250}{360}$

• $\dfrac{250}{360} \times \pi \times 60$

• $130 \cdot 8\ldots$ / 131 cm

12. • $3\cos x - 1 = 0$

• $\cos x = \dfrac{1}{3}$

• $70 \cdot 5$

• $289 \cdot 5$

13. • $x(x-4)$

• $(x-4)(x+5)$

• $\dfrac{x}{x+5}$

14. • $s - ut = \dfrac{1}{2}at^2$

• $2(s - ut) = at^2$

• $a = \dfrac{2(s - ut)}{t^2}$

15. (a) • $130°$

• $\dfrac{50}{\sin CDH} = \dfrac{79}{\sin 130}$

• $\sin CDH = \dfrac{50 \sin 130}{79}$

• $29°$

(b) • 40

• $249 \ [180 + 40 + 29] \ / \ 249°$

16. (a) (i) • $2x + 13$

(ii) • $(2x + 13)(2x + 9) = 4x^2 + 44x + 117$

• $4x^2 + 44x + 117 = 270$
$\Rightarrow 4x^2 + 44x - 153 = 0$

(b) • $\dfrac{-44 \pm \sqrt{44^2 - 4 \times 4 \times (-153)}}{2 \times 4}$

• $\dfrac{-44 \pm \sqrt{4384}}{2 \times 4}$
(stated or implied)

• $2 \cdot 77...$ and $-13 \cdot 77...$

• $2 \cdot 8$ cm

NATIONAL 5 MATHEMATICS 2018

One mark is available for each •. There are no half marks.

Paper 1

1. • $2\dfrac{\cdots}{15} + \dfrac{\cdots}{15}$ **or** $\dfrac{\cdots}{15} + \dfrac{\cdots}{15}$

• $3\dfrac{2}{15}$ or $\dfrac{47}{15}$

2. • $3x^2 - 3x + x - 1$ **or** $2x^2 - 10$

• $3x^2 - 3x + x - 1 + 2x^2 - 10$

• $5x^2 - 2x - 11$

3. • e.g. $\begin{array}{l} 8x + 10y = -6 \\ 30x - 10y = 25 \end{array}$

• values for x and y

• $x = 0 \cdot 5, \ y = -1$

4. • e.g. $\begin{pmatrix} 6 \\ -4 \\ 3 \end{pmatrix} - \begin{pmatrix} 1 \\ 5 \\ 1 \end{pmatrix}$

• $\begin{pmatrix} 5 \\ -9 \\ 2 \end{pmatrix}$

5. • $(x - 3)(x - 8)$

• $(x =) \, 3, \ (x =) \, 8$

6. • $a = 5$

• $b = 4$

7. (a) Method 1

• $\dfrac{6}{4}$ or equivalent

• e.g. $y - 20 = \dfrac{6}{4}(x - 12)$

• $P = \dfrac{3}{2}d + 2$ or equivalent

Method 2

• $\dfrac{6}{4}$ or equivalent

• e.g. $20 = \dfrac{6}{4} \times 12 + c$

• $P = \dfrac{3}{2}d + 2$ or equivalent

(b) • $(£) \, 9 \cdot 50$

8. • −24
   • no real roots
9. • interior angle $= 72 + 72$
   or JKL $= 36$
   • 127
10. • $10^2 + 8^2 - 2 \times 10 \times 8 \times \frac{1}{8}$
    • $XY^2 = 144$
    • $XY = 12$ (cm)
11. • $\frac{9\sqrt{6}}{6}$
    • $\frac{3\sqrt{6}}{2}$
12. • $-0 \cdot 5$
13. • coordinates of B(4, 8, 5)
    • coordinates of C(6, 8, 0)
14. • $y - h = g\sqrt{x}$
    • $\sqrt{x} = \frac{y-h}{g}$
    • $x = \left(\frac{y-h}{g}\right)^2$
15. • $\frac{4}{9}$ **or** p^8
    • $\frac{4}{9}p^8$
16. • −4 **AND** 6
    • (1, −25) **OR** −24
    • turning point = (1, −25) **AND** y-intercept = −24 and consistently annotated parabola.

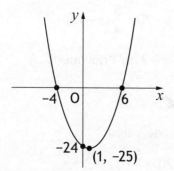

17. **Method 1**
    • $\frac{1}{3} \times 6^2 \times h$ or $\frac{1}{3}Ah = 138$
    • $\frac{1}{3} \times 6^2 \times h = 138$
    • 11·5 (cm)

Method 2
• $\frac{3V}{A}$ or $\frac{V}{\frac{1}{3}A}$
• $\frac{3 \times 138}{6 \times 6}$ or $\frac{138}{\frac{1}{3} \times 6 \times 6}$
• 11·5 (cm)
18. • $\sin x \cos x \frac{\sin x}{\cos x}$
    • $\sin^2 x$
19. (a) (i) • $(x-3)^2 \ldots$
           • $(x-3)^2 - 90$
       (ii) • $x = 3$
    (b) **Method 1**
        • $(x-3)^2 - 90 = 0$
        • $x - 3 = \pm\sqrt{90}$
        • $x = 3 \pm \sqrt{90}$
        • $d = 3, e = 10$ or $3 \pm 3\sqrt{10}$
        Method 2
        • $\frac{6 \pm \sqrt{(-6)^2 - 4 \times 1 \times (-81)}}{2 \times 1}$
        • discriminant = 360 (stated or implied)
        • discriminant = $6\sqrt{10}$
        • $d = 3, e = 10$ or $3 \pm 3\sqrt{10}$

Paper 2
1. • $\times 0 \cdot 98$
   • $125\,000 \times 0 \cdot 98^3$
   • 117 649 (tonnes)
2. **Method 1**
   • $\frac{320}{360}$
   • $\frac{320}{360} \times 2 \times \pi \times 7 \cdot 4$
   • 41(·32…) (cm)
   Method 2
   • $\frac{360}{320}$
   • $2 \times \pi \times 7 \cdot 4 \div \frac{360}{320}$
   • 41(·32…) (cm)

3. • $24^2 + (-12)^2 + 8^2$

   • 28

4. • $3x < 6x - 6 - 12$

   • $-3x < -18$ or $18 < 3x$

   • $x > 6$ or $6 < x$

5. (a) **Method 1**

   • mean = 126

   • $(x - \bar{x})^2 =$ 36, 0, 1, 25, 16, 4

   • $\sqrt{\dfrac{82}{5}}$

   • 4(·049...)

   Method 2

   • 126

   • $\sum x = 756$ and $\sum x^2 = 95338$

   • $\sqrt{\dfrac{95338 - \dfrac{756^2}{6}}{5}}$

   • 4(·049...)

   (b) • e.g. on average the number of customers was higher on Saturday

   • e.g. the number of customers was less varied on Saturday

6. • $5 + 4a = 73$ or $5 + 4 \times 17$

   • $(a =) 17$

7. • $\dfrac{4}{3} \times \pi \times 3 \cdot 2^3$

   • volume = 137·2...

   • 140 (cm³)

8. • $\sin x = \dfrac{1}{7}$

   • 8·2(1...)

   • 171·8 or 171·7(8...)

9. • $\dfrac{20}{\sin 37} = \dfrac{DC}{\sin 105}$

   • $\dfrac{20 \sin 105}{\sin 37}$

   • 32 (·1... cm)

10. • $\overrightarrow{ED} = 2u$ and $\overrightarrow{DC} = \dfrac{1}{2}w$

   • $u - \dfrac{1}{2}w$

11. • $85\% = 9 \cdot 3 \times 10^{11}$

   • $1\% = \dfrac{9 \cdot 3 \times 10^{11}}{85}$

   • $1 \cdot 094... \times 10^{12}$ (km³)

   or 1094117647000 (km³)

12. Method 1

   •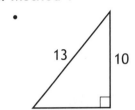

   • $x^2 = 13^2 - 10^2$

   • $x = 8 \cdot 3(...)$

   • width = 21·3(... cm)

   Method 2

   •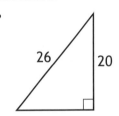

   • $x^2 = 26^2 - 20^2$

   • $x = 16 \cdot 6 (...)$

   • width = 21·3 (... cm)

13. • $\dfrac{10 \cdot 3^2 + 5 \cdot 6^2 - 7 \cdot 2^2}{2 \times 10 \cdot 3 \times 5 \cdot 6}$

   • $\cos YTF = \dfrac{85 \cdot 61}{115 \cdot 36} (= 0 \cdot 742...)$

   • angle YTF = 42(·088...)

   • bearing = 282(·088...)

14. Method 1

   • $-5y = ... + 20$ or $... - 20 = 5y$

   or $\dfrac{2x}{5} - \dfrac{5y}{5} = \dfrac{20}{5}$

   • coordinates = (0, −4)

   Method 2

   • $2 \times 0 - 5y = 20$

   • coordinates = (0, −4)

15. • $\dfrac{n}{n^2-4} \times \dfrac{n-2}{3}$

    • $n^2-4 = (n+2)(n-2)$

    • $\dfrac{n}{3(n+2)}$ or $\dfrac{n}{3n+6}$ s

16. • $40^2 + 40^2$ or $40^2 + 70^2$

    • $\sqrt{40^2 + 40^2 + 70^2}$

    • 90

    • Yes, since $85 < 90$

17. • $\dfrac{1}{2} \times 38 \times 55 \times \sin 75 \,(= 1009 \cdot 39 \ldots)$

    • $\dfrac{75}{360}$

    • $\dfrac{75}{360} \times \pi \times 30^2 \,(= 589 \cdot 04 \ldots)$

    • evidence of area of triangle – area of sector

    • $420\,(\cdot 3 \ldots)$ cm^2

18. (a) **Method 1**

    •^{1} e.g. $\dfrac{24}{16}$ or equivalent

    •^{2} $576 \times \left(\dfrac{24}{16}\right)^3$

    •^{3} $1944 \neq 1125$, so the cartons are not similar

    Method 2

    •^{1} e.g. $\dfrac{16}{24}$ or equivalent

    •^{2} $576 \div \left(\dfrac{16}{24}\right)^3$

    •^{3} $1944 \neq 1125$, so the cartons are not similar

    Method 3

    • e.g. volume scale factor $= \dfrac{1125}{576}$ or equivalent

    • $\sqrt[3]{\dfrac{1125}{576}} \times 16$

    • $20 \neq 24$, so the cartons are not similar

    Method 4

    • e.g. linear scale factor $= \dfrac{24}{16}$ or equivalent

    • $\left(\dfrac{24}{16}\right)^3$ and $\dfrac{1125}{576}$

    • $3 \cdot 375 \neq 1 \cdot 95 \ldots$, so the cartons are not similar

    Method 5

    • e.g. volume scale factor $= \dfrac{1125}{576}$ or equivalent

    • $\sqrt[3]{\dfrac{1125}{576}}$ and $\dfrac{24}{16}$

    • $1 \cdot 25 \neq 1 \cdot 5$, so the cartons are not similar

(b) **Method 1**

    • volume scale factor $= \dfrac{1500}{576}$

    • $\sqrt[3]{\dfrac{1500}{576}} \times 16 = 22\,(\cdot 0 \ldots$ cm$)$

    Method 2

    • volume scale factor $= \dfrac{576}{1500}$

    • $16 \div \sqrt[3]{\dfrac{576}{1500}} = 22\,(\cdot 0 \ldots$ cm$)$

NATIONAL 5 MATHEMATICS 2019

One mark is available for each •. There are no half marks.

Paper 1

1. • $5(-2)^3$ or equivalent
    • -40

2. • $\dfrac{3}{8} \times \dfrac{12}{7}$
    • $\dfrac{9}{14}$

3. • evidence of any 3 correct terms
      e.g. $2x^3 - 7x^2 - 3x$
    • $2x^3 - 7x^2 - 3x + 10x^2 - 35x - 15$
    • $2x^3 + 3x^2 - 38x - 15$

4. **Method 1**
    • $\dfrac{240}{360}$ or equivalent
    • $\dfrac{240}{360} \times 3 \cdot 14 \times 60$
    • $125 \cdot 6$ (cm)

    Method 2
    • $\dfrac{240}{360}$ or equivalent
    • $\dfrac{240}{360} = \dfrac{\text{arc}}{3 \cdot 14 \times 60}$
    • $125 \cdot 6$ (cm)

5. (a) • median $= 5$
        • quartiles $3 \cdot 5$ and 8
        • SIQR $= \dfrac{1}{2}(8 - 3 \cdot 5) = 2 \cdot 25$
    (b) • e.g. On average, temperatures in Grantford are lower.
        • e.g. Temperatures in Grantford are less consistent.

6. (a) **Method 1**
        • gradient $= -\dfrac{6}{2}$ or equivalent
        • e.g. $y - 8 = -\dfrac{6}{2}(x - 3 \cdot 5)$
        • e.g. $F = -3E + 18 \cdot 5$

    Method 2
        • gradient $= -\dfrac{6}{2}$ or equivalent
        • e.g. $8 = -\dfrac{6}{2} \times 3 \cdot 5 + c$
        • e.g. $F = -3E + 18 \cdot 5$

    (b) • $15 \cdot 2$ (km/l)

7. **Method 1**
    • $2A = h(x + y)$
    • $\dfrac{2A}{h} = x + y$
    • $x = \dfrac{2A}{h} - y$

    Method 2
    • $2A = h(x + y)$
    • $2A - hy = hx$
    • $x = \dfrac{2A - hy}{h}$

8. (a) • e.g. $7c + 3g = 215$
    (b) • e.g. $5c + 4g = 200$
    (c) • e.g. $\begin{array}{l} 28c + 12g = 860 \\ 15c + 12g = 600 \end{array}$
        or $\begin{array}{l} 35c + 15g = 1075 \\ 35c + 28g = 1400 \end{array}$
        • $c = 20$ or $g = 25$
        • $g = 25$ or $c = 20$
        • cement $= 20$kg, gravel $= 25$kg

9. (a) • $x = 4$
    (b) (i) • -4
        (ii) • 20

10. (a) • $\begin{pmatrix} 5 \\ 4 \end{pmatrix}$
    (b) • $\dfrac{1}{2}\overrightarrow{PR} + \overrightarrow{RQ}$ or $\dfrac{1}{2}\begin{pmatrix} 6 \\ -4 \end{pmatrix} + \begin{pmatrix} -1 \\ 8 \end{pmatrix}$

        OR $\dfrac{1}{2}\overrightarrow{RP} + \overrightarrow{PQ}$ or $\dfrac{1}{2}\begin{pmatrix} -6 \\ 4 \end{pmatrix} + \begin{pmatrix} 5 \\ 4 \end{pmatrix}$

        • $\begin{pmatrix} 2 \\ 6 \end{pmatrix}$

11. • AOB $= 72$
    • FOB $= 108$ or ABO $= 54$
    • OFB $= 36$

12. **Method 1**
    • $\dfrac{\sqrt{2}\sqrt{40}}{40}$ or $\dfrac{\sqrt{80}}{40}$
    • $\dfrac{4\sqrt{5}}{40}$
    • $\dfrac{\sqrt{5}}{10}$

Method 2

- $\dfrac{\sqrt{2}}{2\sqrt{10}}$ or $\dfrac{\cdots}{2\sqrt{10}}$

- $\dfrac{\sqrt{2}\sqrt{10}}{20}$ or $\dfrac{\sqrt{20}}{20}$

- $\dfrac{\sqrt{5}}{10}$

Method 3

- $\dfrac{1}{\sqrt{20}}$

- $\dfrac{1}{2\sqrt{5}}$

- $\dfrac{\sqrt{5}}{10}$

13. • $(135,\ldots)$

    • $(\ldots,-3)$

14. **Method 1**

    • $5x - 10 = 6 - 2x$ or equivalent

    • $7x = 16$

    • $x = \dfrac{16}{7}$

Method 2

    • $\dfrac{7x-6}{10} = 1$ or equivalent

    • $7x = 16$

    • $x = \dfrac{16}{7}$

15. (a) • $(12 \times 2 - 5 \times 2^2 =)\ 4\,(\text{m})$

    (b) • $12t - 5t^2 = -17$

       • e.g. $5t^2 - 12t - 17 = 0$

       • $(5t - 17)(t + 1)\ (= 0)$

       • $(t =)\ \dfrac{17}{5}$ (seconds) or equivalent

Paper 2

1. • $\times 1\cdot 15$

   • $80\,000 \times 1\cdot 15^3$

   • $121\,670$

2. • $6^2 + 27^2 + (-18)^2$

   • 33

3. • $\dfrac{1}{2} \times 45 \times 70 \times \sin 129$

   • $1224(\cdot 004\ldots)(\text{cm}^2)$

4. • $0\cdot 08 \times 3\cdot 6 \times 10^{-6}$ or equivalent

   • $2\cdot 88 \times 10^{-7}\,(\text{kg})$

5. • $A(3,0,0)$

   • $B(3,3,8)$

6. • $\dfrac{-9 \pm \sqrt{9^2 - 4 \times 3 \times (-2)}}{2 \times 3}$

   • $9^2 - 4 \times 3 \times (-2) = 105$

   • $-3\cdot 2,\ 0\cdot 2$

7. • $(\cos Z =)\dfrac{7\cdot 2^2 + 8\cdot 5^2 - 6\cdot 3^2}{2 \times 7\cdot 2 \times 8\cdot 5}$

   • $(\cos Z =)\dfrac{84\cdot 4}{122\cdot 4}\left(=\dfrac{211}{306} = 0\cdot 689\ldots\right)$

   • $(Z =)46\cdot 406\ldots$

8. • sphere $= \dfrac{4}{3} \times \pi \times 12^3$

     cylinder $= \pi \times 12^2 \times 58$

     total $= \dfrac{1}{2} \times \dfrac{4}{3} \times \pi \times 12^3 + \pi \times 12^2 \times 58$

   • $(3619\cdot 1\ldots + 26238\cdot 5\ldots) = 29\,857\cdot\ldots$

   • $29\,900\ \text{cm}^3$

9. • $102\cdot 5(\%) = 977\cdot 85$

   • $977\cdot 85 \div 102\cdot 5$ or equivalent

   • $(£)23\cdot 85$

10. • $(x + 5)^2\ldots$

    • $(x\ldots 5) - 40$

11. **Method 1**

    • e.g. $600^2 + 250^2$ and 650^2

    • $600^2 + 250^2 = 422\,500$ and $650^2 = 422\,500$

    • $600^2 + 250^2 = 650^2$

    • Yes, as angle is a right angle.

Method 2

    • $(\cos B =)\dfrac{600^2 + 250^2 - 650^2}{2 \times 600 \times 250}$

    • $(\cos B =)0$

    • $(B =)90$ [stated explicitly]

    • Yes, as angle is a right angle.

12. (a) Method 1

- $\dfrac{30}{50}$

- $2750 \times \left(\dfrac{30}{50}\right)^2$

- 990 (cm²)

Method 2

- $\dfrac{50}{30}$

- $2750 \div \left(\dfrac{50}{30}\right)^2$

- 990 (cm²)

Method 3
[Combination of (b) and (a)]

- • • $126(\cdot 05\ldots)$

- $\dfrac{126(\cdot 05\ldots)}{360}$

- $\dfrac{126(\cdot 05\ldots)}{360} \times \pi \times 30^2$

- 990 (cm²)

(b) Method 1

- $\dfrac{\text{angle}}{360} \times \pi \times 50^2$

- $\dfrac{2750 \times 360}{\pi \times 50^2}$

- $126(\cdot 50\ldots)$

Method 2

- $\dfrac{2750}{\pi \times 50^2}$ $(= 0\cdot 35\ldots)$

- $\dfrac{2750 \times 360}{\pi \times 50^2}$

- $126(\cdot 05\ldots)$

13.

- $\dfrac{4p^2-9}{4p-6}$ or $\dfrac{9-4p^2}{6-4p}$

- $(2p+3)(2p-3)$
  or
  $(3+2p)(3-2p)$

- $\dfrac{(2p+3)(2p-3)}{2(2p-3)} = \dfrac{2p+3}{2}$
  or
  $\dfrac{(3+2p)(3-2p)}{2(3-2p)} = \dfrac{3+2p}{2}$

14.

- $\cos x = -\dfrac{1}{5}$ or equivalent

- $101\cdot 5(3\ldots)$

- $258\cdot 4(6\ldots)$

15.

- denominator $= (x-2)(x+5)$

- numerator $= 4(x+5) - 3(x-2)$

- $\dfrac{x+26}{(x-2)(x+5)}$

16.

- e.g. $a^4 \times 3a = 3a^5$

- numerator $= 3a^5$
  denominator $= a^{\frac{1}{2}}$

- $3a^{\frac{9}{2}}$

17.

- $\sin^2 x + \sin x \cos x + \cos x \sin x + \cos^2 x$

- $1 + 2\sin x \cos x$

18.

- $r^2 = 7\cdot 5^2 + 7\cdot 5^2$

- $r = 10\cdot 6\ldots$

- CD $= 25\cdot 6\ldots$ (cm)

19. Method 1

- $\dfrac{BK}{\sin 34} = \dfrac{350}{\sin 94}$

- $BK = \dfrac{350\sin 34}{\sin 94}$

- $196(\cdot 195\ldots)$

- $\sin 52 = \dfrac{h}{196}$ or $\dfrac{h}{\sin 52} = \dfrac{196}{\sin 90}$

- $154\cdot 6$ (m)

Method 2

- $\dfrac{BM}{\sin 52} = \dfrac{350}{\sin 94}$

- $BM = \dfrac{350\sin 52}{\sin 94}$

- $276(\cdot 477\ldots)$

- $\sin 34 = \dfrac{h}{276}$ or $\dfrac{h}{\sin 34} = \dfrac{276}{\sin 90}$

- $154\cdot 6$ (m)

Acknowledgements

Permission has been sought from all relevant copyright holders and Hodder Gibson is grateful for the use of the following:

Image © Dotted Yeti/Shutterstock.com (2018 Paper 2 page 12);
Image © TeddyGraphics/Shutterstock.com (2018 Paper 2 page 12);
Image © Richard Peterson/Shutterstock.com (2018 Paper 2 page 18);
Image © Steinar/Shutterstock.com (2019 Paper 2 page 10).